Buckminster Brown

Cases in Orthopedic Surgery

Read before the Massachusetts Medical Society at its annual meeting, June 3,

1868

Buckminster Brown

Cases in Orthopedic Surgery
Read before the Massachusetts Medical Society at its annual meeting, June 3, 1868

ISBN/EAN: 9783743407183

Manufactured in Europe, USA, Canada, Australia, Japa

Cover: Foto ©berggeist007 / pixelio.de

Manufactured and distributed by brebook publishing software
(www.brebook.com)

Buckminster Brown

Cases in Orthopedic Surgery

CASES

IN

ORTHOPEDIC SURGERY:

READ BEFORE

The Massachusetts Medical Society,

AT ITS

ANNUAL MEETING, JUNE 3, 1868.

By BUCKMINSTER BROWN, M.D.,

Fellow of the Society, Member of the Boston Society for Medical Improvement, etc. etc.

WITH PHOTOGRAPHIC ILLUSTRATIONS OF THE CASES PRESENTED.

BOSTON:

DAVID CLAPP & SON, 334 WASHINGTON ST.

1868.

ORTHOPEDIC SURGERY.

MEMBERS of the Profession, both those residing in the city and those coming from a distance, are frequently reminding me that cases showing the results which can now be attained by combined operative and mechanical surgery, in the treatment of deformities, possess a great interest for the general practitioner. Acting upon these suggestions, it is proposed at this time to state, in few words, the history of several cases, most of which have recently come under my observation, and are brought forward as examples of some of the various classes into which this branch of surgery is divided. The better to elucidate the subject, casts or photographs will be shown, before and after treatment; and, in two or three instances, I am enabled to present the patient to the Society for examination.

CASE I. Casts Nos. 1 and 2. Photographs 1 and 2.

TALIPES.—The first case to which I will draw attention is the one from which this cast and this photograph were taken. (See Case I., Plate I., figure 1.) A boy born with such a distortion of the leg and foot that the great toe was turned up against the side of the knee, and, when the child was awake, was in close contact with the internal condyle of the femur. The tibialis anticus and posticus muscles were strongly contracted—structurally shortened. The treatment consisted in the division of the tendons of these muscles, and in the use of a variety of apparatus, employing sometimes the spring and sometimes the screw power. By these means the leg and foot were gradually brought into a normal position. By the time, however, this result was somewhat more than half accomplished, the tendons, growing more rapidly than the bones, had united, and again presented an obstacle to further improvement. These were re-divided, and, in about six months, the result was as shown in the second cast. (See Case I., Plate I., figure 2.) The second photograph gives a correct idea of the foot when it was nearly straight. When I last saw the child, he walked on the sole of his foot.

CASE II.   Casts Nos. 1, 2, 3, 4.

This is a case of paralytic calcaneo-valgus, the result of spina bifida. (See Case II., Plate I., figures 3 and 5.) The patient is a girl eleven years old. It is evident that in the left foot the articular facet of the astragalus, instead of being applied against the internal malleolus, does not enter into the composition of the ankle-joint, but, with its rounded internal face, and with the scaphoid, forms the projection on the inside of the foot. It will be seen that the front part of the foot is higher than the heel, therefore it is calcaneo-valgus; yet, in reality, from the displacement of the calcaneum, the origin and insertion of the gastrocnemius are approximated. Thus is presented the somewhat anomalous state of the parts in which, although treating a case of calcaneus, instead of desiring to shorten the tendo-Achillis, we are obliged to increase its length before the foot can be replaced. The tendons divided in the left foot were the three peronei, the extensor longus-communis, extensor pollicis-pedis, the tibialis anticus and the tendo-Achillis. The same tendons were divided in the right foot, with the exception of the tibialis anticus and the tendo-Achillis. I here present the models of the feet, as they were before treatment, and four months after the commencement of treatment. (See Case II., Plate I., figures 3, 4, 5, 6.) The child is able to walk on the soles of the feet.

CASE III.　Casts 1 and 2.　(Patient present.)

There are but few cases in which the patients are so situated as to appear on such an occasion as this. Either they live at too great a distance, or they dislike to be presented. This boy, however, willingly comes forward. He well illustrates the legitimate results of the combination of operative and mechanical surgery. He is nine years of age. He had congenital varus of the right foot. Continued use of the foot had converted it into what has been termed varo-dorsalis. The foot had been operated upon some years previous to

coming under my care. I divided the tibialis anticus, **plantar
fascia** and tendo-Achillis. (See Case III., Plate II., figures
1 and 2.)

CASE IV.   Cast No. 1, and Patient.

I have here a cast of one foot from a case of double talipes varus. The feet were alike. (The patient will stand
upon the table, or walk around, that gentlemen **who wish**
may examine her feet.) (See Case IV., Plate II., **figures 3
and 4.**)

CASE , V.   Casts Nos. 1 and 2, and Patient.

This cast (see Case V., Plate II., figure 5) speaks for
itself. The child is three and a half years old. She had
paralytic varus.   The tendo-Achillis, tibialis anticus and
posticus, and **extensor** longus pollicis pedis were rigidly contracted. These tendons were divided **in February,** 1868.
In **twenty** days after the operation, the **paralyzed** muscles,
no longer kept fully extended by their contracted antagonists,
completely recovered their power, and the child was able to
flex and abduct the foot. A cure was effected in two months.
(See Case V., Plate II., figure 6.)   The child is now under
treatment for contracted knee, arising from the same cause.
The biceps flexor cruris has been recently divided.

Paralysis is rarely the cause of congenital varus. Non-congenital varus, however, frequently arises from paralysis **of a**
single muscle or of a set of muscles. On the other hand, the

etiology of both congenital and non-congenital *valgus* may so constantly be traced to debility of muscles and ligaments—amounting, in the majority of cases, to complete paralysis—as almost to form the rule in this class of cases. The return of power to the paralyzed muscles I have frequently observed after division of the healthy muscles, which are structurally shortened in consequence of the normal balance of power having been destroyed.

CASE VI.   Casts Nos. 1 and 2.

Within a year or two, much has been said and written in regard to the cure of talipes without tenotomy. This case may be cited (one among several that could be referred to) as a fair instance of the result of such attempts. (See Case VI., Plate III., figure 1.) The child from whom the cast was taken was born with double talipes varus. A few days after birth the treatment by apparatus was commenced, and was continued two years. For three months the patient was visited daily by the attending surgeon. The result, after two years, was a failure, as is shown in the first cast, taken when he came under my treatment. The second cast was taken three months afterwards. (See Case VI., Plate III., figure 2.)

CASE VII.   Casts Nos. 1 and 2.

The cast I have in my hand (see Case VII., Plate III., figure 3), represents a case of varus, interesting from the fact

2

that a somewhat similar attempt to the preceding had been made to cure the foot. It has been said the hand of the mother or nurse is in truth the best apparatus. In this case the mother, instructed by her physician, had devoted herself to the task. She had held the foot in her hands, on the stretch towards a straight line, **four hours a day for three months.** Flexion was impossible from any force that could be applied to it. That her labor was thrown away is shown in this first cast taken when the patient was brought to Boston. This second cast shows the foot after tenotomy and subsequent treatment. (See Case VII., Plate III., fig. 4.)

CASE VIII. Casts Nos. 1 and 2.

I have cited cases showing the nugatory effects of protracted mechanical treatment without operation. Here are a couple of casts, not remarkable in themselves, but interesting as examples of numerous cases exhibiting the same or worse results from the opposite mode of treatment, viz. :—too much surgery without appropriate after-treatment. The lad, from whom this model was taken, had been operated upon six times by a distinguished New York surgeon. The tendo-Achillis was divided three times. After five years treatment the foot was as malformed as at first. (See Case VIII., Plate III., figure 5.) The second model was taken after the boy had been in Boston three months. (See Case VIII., Plate III., figure 6.)

CASE IX.   Casts Nos. 1 and 2.

These casts furnish another instance of the fact just alluded to.   The boy had double varus, third degree.   He had been operated upon ten or twelve times; and had likewise been under treatment five years by surgeons in New York and at the West, with the disastrous result seen in the first cast.   (See Case IX., Plate IV., figure 1.)   The second cast shows the feet (they were alike) when he left Boston. (See Case IX., Plate IV., figure 2.)

CASES X. and XI.   Casts.

In order to make the series more complete, there are on the table, one sample of talipes equinus before and after treatment (See Case X., Plate IV., figures 3 and 4); also casts of a case of varus, treated several years since, introduced simply to show that the growth and strength of the members are not diminished by somewhat extensive tenotomy. The case was one of extreme double varus.   The tendo-Achillis in each foot was twice divided, also the tibialis posticus and flexor longus pollicis pedis.   The boy was treated and cured, when eight years of age.   The second cast was taken twelve years afterwards.   (See Case II., Plate V., figures 1 and 2.)   These years the lad had passed chiefly at sea, doing duty as a sailor.   He has since become master of a vessel, and states that he has never experienced the slightest inconvenience from his feet.

CASE XII.   Casts Nos. 1 and 2.

This is an example of a case of genu-varum or bow-legs. (See Case XII., Plate **V.**, figure 3.)   Both legs were similarly affected.   It is curious to notice, that, although the legs were tightly strapped on the **convex side** for months, yet the healthy growth and development **were** not impeded, as is well **shown** in the second **cast.**   (See Case XII., Plate V., figure **4.**)   This will **be found to be the rule in** all cases where **the apparatus is so applied as not** materially to interfere with the circulation.

CASE XIII.   Photographs Nos. 1 and 2.

This photograph (see Case XIII., Plate VI., figure 1) represents the legs of a little girl as they were when she came under treatment.   It was a bad case of genu-valgum of the right leg, and genu-varum of the left.   The result, as shown in the accompanying photograph, was attained by apparatus without tenotomy.   (See Case XIII., Plate VI., figure 2.)

CASE XIV.   Photographs Nos. 1 and 2.

These photographs were taken from another case, of a similar nature to the preceding, before and after treatment. In this case, also, no operation was required.   The distortion, in both instances, was caused by malformation of the joints, uncomplicated by muscular contraction.*

---

* The photographs of Case XIV. are not copied for publication, as the case resembles that represented on Plate VI.

LATERAL CURVATURE OF THE SPINE, or, according to the latest and best authority,* "Rotato-Lateral Curvature,", in its advanced stages, is one of the most discouraging affections with which we have to deal. Much, however, can be accomplished by patience and perseverance. In spinal curvature, as in most other cases pertaining to this branch of surgery, frequent variation of the treatment, and, where apparatus is employed, a frequent change in the appliances, is required.† They should be modified according to the exigencies of the case, adapting the means employed to the changes in the form as the cure proceeds. It may be interesting to mention an extreme case of this complaint which has recently come under my observation.

CASE XV.

Miss ——, aged 31. Has had curvature of the spine from childhood. Her body, from neck to hips, has gradually shortened. For this there is a partial compensation in the greatly increased antero-posterior diameter of the chest. On examination, I found the crest of the ilium, on the left side, to be two inches from the axilla. On the right, the distance is two and three-fourths inches. In fact these bones are lodged directly beneath the shoulders. The os pubis is

---

* On Spinal Weakness and Spinal Curvature; its early Recognition and Treatment. By W. J. Little, M.D. London, 1868.

† This rule applies, with especial force, to talipes. In every species of club-foot, excepting where the twist is very slight, from two to six, or even more varieties of apparatus are often required to make a perfect foot.

three and one-half inches from the sternum. Relief, by an accurately adjusted support, was the only treatment admissible. Such extraordinary cases are rarely met with.

CASE XVI.   Photographs **1 and 2.**

Here is a photograph of the **back of a boy from Lawrence, Kanzas.** (See Case XVI., Plate **VII.,** figure **1.**) He had severe lateral curvature. The left hip was very prominent. The trunk, above the hips, **was thrown so far to the right, that the centre** of the occiput was on a line **with the right leg;** consequently, in standing, the weight of the **body was** sustained by this **leg. The** right scapula and **ribs projected, and the left scapula sank into** the hollow **formed by the curve. This unequal distribution** of the weight **of the body had produced an inward inclination** of the left **knee. The second** photograph (See **Case XVI., Plate VII.,** figure 2) shows the **state of** the **spine some months since.** It is now still further improved. **The left shoulder,** formerly much below the level of **the** right, is now **the higher.** This **will** rectify itself. The knee was cured by **proper apparatus.**

CARIES OF THE CERVICAL **VERTEBRÆ,** compared with the **same** affection as **it attacks** other regions of the spinal column, is a rare disease. **Some** years since **I** published **an account of a case of caries of** the upper bones of the **neck, remarkable in many points of view,** which terminated fatally. The atlas, axis and base **of** the cranium were eroded, and death **was caused** by fracture of the odontoid process. The pathological appearances were minutely described. About the same time two or three similar instances presented themselves, which were also fatal. Since then I have treated other cases of cervical caries which have had a more favorable termination. There are present, to-day, two children who have been sufferers from this disease.

CASE XVII.   (Patient.)

This little **girl, when** I first saw her, **eleven months since,** had **lost all voluntary** power below her **neck.   She could speak in a whisper.   The only muscles not paralyzed were those connected with the eyes and** mouth.   **She had been in this** state some months.   There was swelling and prominence of the lower **cervical** vertebræ.   **In February, 1866, an ab-**scess formed in the neck, which continued discharging, **at in-**tervals, for twelve months.   **She then began to lose the use** of her right arm and leg.   The paralysis extended, involving both arms **and legs, with inability to move the** head. For a time **the bladder** was implicated, and **the** use of **a** catheter was required.   She had paroxysms **of** severe **pain** in the diseased bones.   **The child, as you see, is now well and without deformity.   There is scarcely a trace of the affection** remaining.

I have recently been informed that the elder sister of this patient **died of caries of the dorsal** vertebræ, after **having been** paralyzed **three years.**

CASE XVIII. (Patient.)

This boy had the same disease in about the same situation. The symptoms, also, were very nearly similar, but had been of longer duration when he came under my observation. There was complete paralysis of all the voluntary muscles below the mouth. He had been unable, for months, to move his head, or to bend a finger or a toe, or to speak above a whisper. Severe pain was produced if any attempt was made to bend his fingers, wrists, knees or ankles. The joints were stiff. He had incontinence of urine. The respiratory muscles acted imperfectly, and his breathing was labored. His countenance expressed suffering, and his manner of rolling his eyes, to compensate for inability to move his head, gave him a very singular appearance. The paralysis commenced about ten months previous to his being placed under my care. The treatment consisted, in the first place, of mechanical support to retain the head in one position. The apparatus was a spring collar, resting on the clavicles and shoulders, with branches running down the back, and secured by a belt. Passive exercise of all the joints was perseveringly employed. Friction, electro-magnetism, the pyro-phosphate of iron, and cod liver oil were important adjuncts. Chloroform was given internally, to relieve pain. Power of motion returned first to his fingers, and gradually extended, and in three months he began to walk.

In this case it is interesting to notice that the efforts of nature to cure the disease have exceeded the necessity; and there has been a great amount of ossific matter thrown out around the bones, producing considerable deformity of the

neck.   An abscess formed and opened spontaneously.   The
boy is now able to walk long distances, carrying bundles
and going on errands.

CASE XIX.   (Patient.)

HIP DISEASE.—It is impossible, in the brief time to which
these papers are necessarily limited, to give more than a very
imperfect sketch of the various diseases, and their effects in
deranging and distorting the human frame, which receive their
proper classification in the branch of surgery we are now
considering.   The cases already brought forward are of practical importance.   Those last introduced, all will acknow-
ledge, are eminently so.   They are instructive instances of
the recuperative powers of nature, aided and guided by art;
and teach us how much these may be relied upon even in
cases which appear utterly hopeless.   I should be glad, if
time permitted, to draw your attention to the several varieties of hip disease, referring to the diverse, and sometimes
almost opposite modes of treatment appropriate to the different cases and to the different stages of the same case.
The interest attached to these would be increased if the patients, showing in their persons the results, could be presented to you as in some of the preceding cases.   My limits
allow me, at this session, to bring forward one patient only.
He suffered from morbus coxarius from August, 1866, to
March, 1867.   When first seen by me, in October, 1866, he
could not bear the slightest touch in the neighborhood of the

3

left hip, and had severe pain in this joint and in the knee. He had frequent startings in the night, waking and screaming with pain. The patient was seen by Dr. J. Mason Warren, and other surgeons, during the early months of the disease. The treatment was directed, in the first place, to relieving the symptoms of acute action within the joint. It has been much the fashion of late, for surgeons treating hip diseases, and those analogous, of the spine, to discard, as old fashioned, all counter-irritants and antiphlogistic remedies. Let me urge them not to do this in every case. There are certain varieties and phases of these diseases in which there are no other means of relief possible. Mr. Pott was not so utterly mistaken as many in these days would have us believe. There are cases of disease of the hip or spine in which a modification of his treatment is of incalculable benefit, and it is only on account of its indiscriminate employment in all varieties of these complaints, in many of which disappointment has attended its application, that it has fallen into discredit.

In the earlier stages of some species of hip disease, for example, I have too frequently seen the severe pain, the extreme tenderness of the joint—where the slightest jar is agony—the nocturnal startings and spasms, and the pain in the knee, removed, after having existed for months, by flying blisters, or by an issue, preceded, if the state and history of the patient render it advisable, by slight local blood-letting, to have a doubt left in my mind in regard to the importance of these remedies. The relief is often immediate; neither extension, nor rest, nor internal remedies will have the slightest effect in such cases, without the aid of local applications in some one or more of the forms which experience has taught us are most beneficial. Quiet nights and comfortable days were the immediate results of this course in the case now under consideration. A hip-splint was applied, complete rest enjoined, and slight extension was used. The apparatus employed had especial reference to the prevention of contraction or permanent displacement at the joint, one of the most frequent and unfortunate sequelæ of this disease, to obviate which requires the exercise of the utmost caution. The boy has been, for more than a year, in as perfect health as you now see him. There is not the slightest limp. That the hips are alike in appearance and perfectly normal in action, will be acknowledged by those who will examine the patient.

The importance of attention to position in hip complaint cannot be too strongly insisted upon. It may be useful, in this connection, to refer to a case which, probably, has not its counterpart upon record. The patient was a young girl, thirteen years of age, who was brought to me from a distance, a few years since. She had suffered, for many months, from double hip disease. The complaint had gone through its several stages, and had terminated in anchylosis. From malposition, during the acute periods, both thighs had become permanently fixed at right angles with the sides of the body, on a line with the axillæ, and parallel with the arms when stretched in such a manner as to afford the fullest expansion to the chest. In sitting, the lower limbs projected over the sides of the chair. The head of each femur was joined to the acetabulum by a solid, bony union. The case was irremediable. A greater misfortune can scarcely be imagined. Double excision at the hip joint was a procedure maturely considered, but decided not to be advisable under the circumstances.

TORTICOLLIS, a less common affection than any of the preceding, has some curious features which **are** worthy of notice. Its causes are various. Frequently it arises from contraction, congenital or non-congenital, of one or both branches of the sterno-cleido-mastoideus muscle, sometimes combined with a similar affection of the trapezius or scaleni. Spasm, **permanent** or intermittent, the cicatrices of burns, and paralysis, may give rise to this affection. Other instances of wry neck originate in rheumatic inflammation of one or all of the muscles just named. I have found the trapezius condensed into a firm, indurated tissue, apparently as unyielding as ligament. I have also seen very serious distortion arise from rheumatism attacking the inter-vertebral substance between two or more of the cervical vertebræ. This disease will produce swelling and permanent thickening of the ligamentous tissue on one side, the bone, perhaps, being implicated, while the muscles are not at all, or but slightly involved. The peculiar, characteristic, rotatory twist, in these instances, is less observable than in other varieties of torticollis. Those cases originating in the causes first mentioned are generally incurable without division of the offending muscles. The last named may, often, be completely relieved by apparatus and appropriate remedies. An appliance which shall fix the head, and enable us to act upon it steadily and firmly, has been a desideratum in surgery, not only for the treatment of the complaint we are now considering, but, also, when dealing with the deformities arising as a sequence of burns and from other causes. I have seen but one apparatus that does this effectually. It was invented by Dr. John B. Brown, some years since. This instrument answers every indication in torticollis, and, for accomplishing the object desired, is nearly perfect. We have, in this affection, to contend, first, with the sideward inclination of the head, which sometimes almost touches the shoulder; secondly, with the rotation, by which the face is turned towards the opposite shoulder; thirdly, with the tendency to stoop, or posterior curvature of the dorsal vertebræ; and, fourthly, with the lateral curvature, which is the inevitable result of the disturbance of the equilibrium above. The apparatus referred to, consists of a padded steel belt, which firmly grasps the pelvis. . From the centre, opposite the sacrum, arises a strong steel upright, terminating in a steel skull-cap, which encircles the head, with a tongue, projecting obliquely downwards and forwards, to press upon the superior and inferior maxillary bones. There is a crutch, on one side, to balance the instrument and to support the depressed shoulder. A broad belt, also of steel, attached to the back upright, embraces the body below the axillæ, and buckles in front. About two inches below the cap, upon the posterior standard and opposite the cervical vertebræ, is a circular ratchet-wheel which acts in such a way as to rotate the head ; below this, another, working in a different direction, tilts the head towards either shoulder.   A third ratchet-wheel, opposite the middle dorsal vertebræ, acts antero-posteriorly, and corrects the stoop which is a frequent accompaniment of torticollis.   A fourth is placed near the insertion of the standard to the pelvic belt.   This turns from side to side, and, by means of the upper belt, below the shoulders, unbends the lateral curve of the spine, which, as has been stated above, is always present in a confirmed case of wry neck.

For the milder varieties of this affection, where less power is required, I have contrived a less complicated but very efficient instrument.   This is a double spring stock to sustain the head, from which two pieces of steel, about a foot long and half an inch wide, extend down each side of the spine and are secured to the waist by a leather belt.   There is a check for the chin, and a spring against the occiput, by which the head is turned and retained in a position the reverse of that towards which it is abnormally inclined.

Another of the sequelæ of torticollis is the series of remarkable twists which gradually occur in the facial lineaments.   The physiognomy becomes characteristic.   The eyes, nose, mouth, and even the eye-brows, endeavor to adapt themselves to the one-sided position of the head.   There is a persistent, involuntary effort made, by the muscles, to compensate for this obliquity and to restore the normal, relative position of the features.   This, in time, produces a very peculiar appearance of the countenance, which is pathognomonic of the complaint.

CASE XX. Photographs Nos. 1 and 2.

The last case, of which I have to speak this morning, is that of a girl twenty years of age, whose situation before treatment is shown in this photograph. (See Case XX., Plate VIII., figure 1.) When eleven years old, while at play, she was thrown from a height of sixteen feet, by the caving in of an embankment, the lumbar and sacral spine striking upon a large stone. The fall produced insensibility for a few moments. She then recovered and went to school. She continued her usual avocations for five or six weeks, growing, daily, more and more feeble. She was then attacked with agonizing pain in the lumbar region, followed by complete loss of sensation and motion below the hips. The thighs and legs gradually contracted, until the left knee was forced against, and under, the right thigh, and the thigh was drawn up to an acute angle with the body, and twisted to the right. These parts were in such close contact that it was with difficulty I forced the knee from under the thigh where it had lain for years. The patient had extreme lateral curvature, with excessive incurvation of the lumbar vertebræ. The first photograph was taken nine years after the accident. By means of subcutaneous division of tendons in the groins, popliteal regions and in the feet, followed by mechanical appliances, together with a carefully adjusted spinal apparatus, the girl was in three months straightened out as seen in the second representation. (See Case XX., Plate VIII., figure 2.)

There are other photographs and models upon the table, for examination, by any gentlemen who may feel interested.

PLATES.

THE figures in the accompanying Plates are photographic representations of most of the cases described in the preceding paper. They are copied with an accuracy only attainable by that wonderful art which permits the subject to stamp its own image.

Each one is the type of a class, or is illustrative of practical facts referred to in the text, and is indicated by numerals under the appropriate heads.

PLATE I.

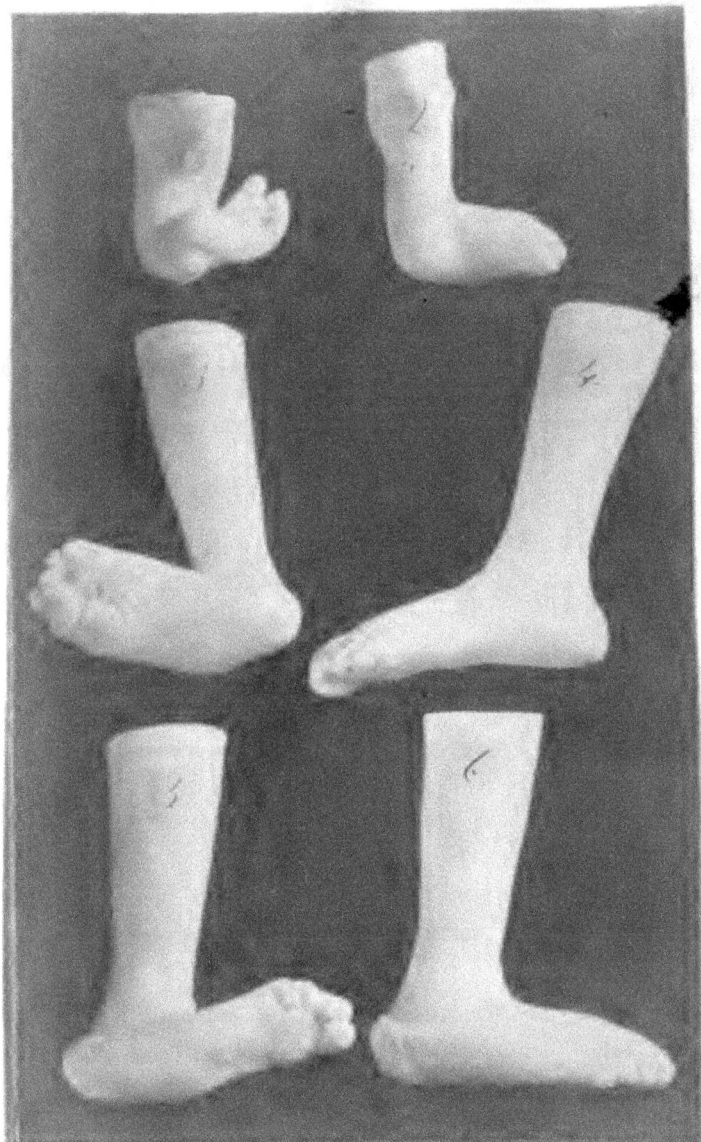

CASE I.

Fig. 1. Before treatment.   Fig. 2. After treatment.

CASE II.

Fig. 3. Left foot before treatment.   Fig. 4. Left foot after treatment.
Fig. 5. Right foot before treatment.   Fig. 6. Right foot after treatment.

PLATE II.

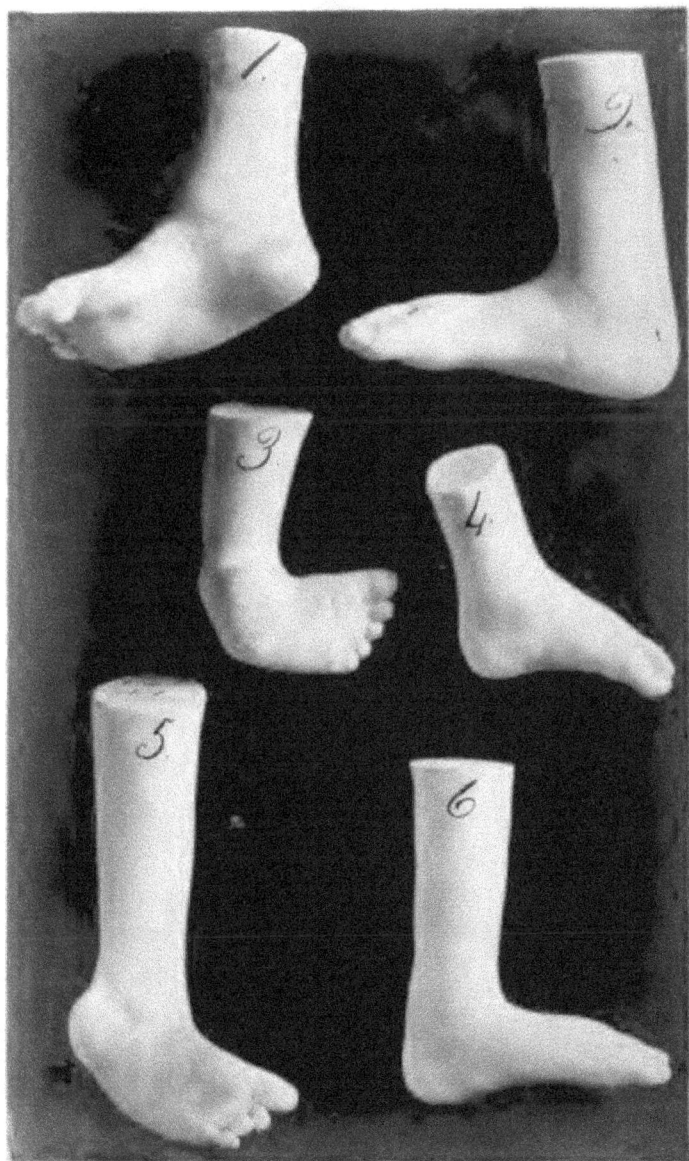

CASE III.

Fig. 1. Before treatment.   Fig. 2. After treatment.

CASE IV.

Fig. 3. Before treatment.   Fig. 4. After treatment.

CASE V.

Fig. 5. Before treatment.   Fig. 6. After treatment.

J. T. PAWS.   PHOTO.

PLATE III.

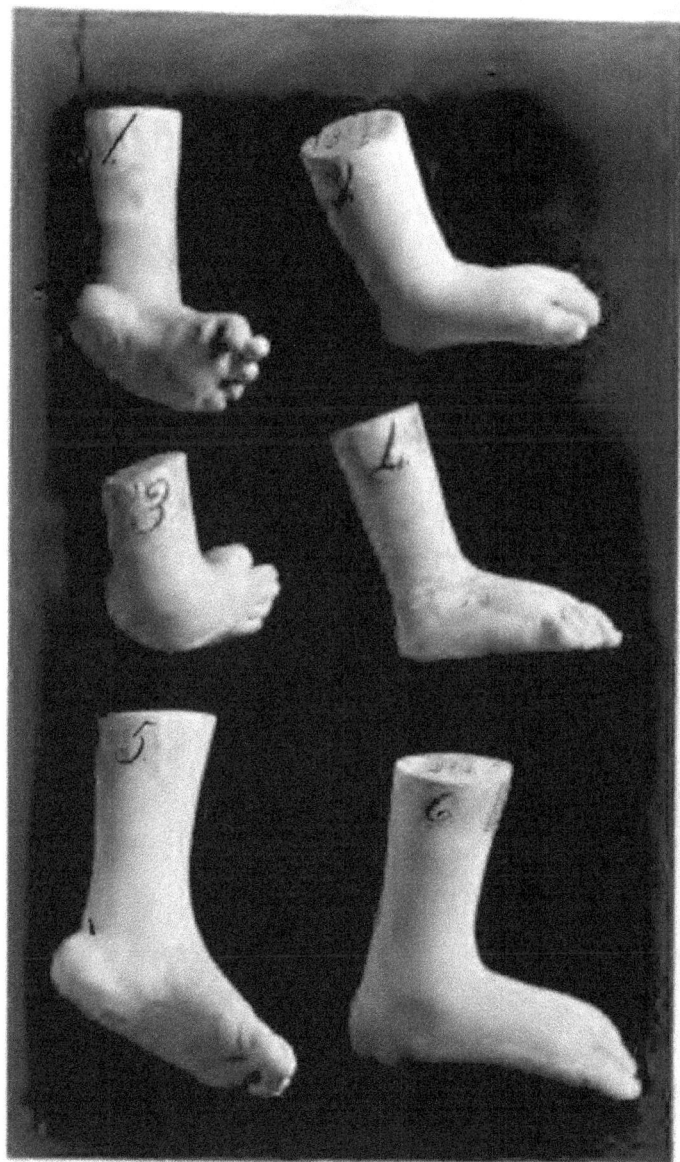

CASE VI.

Fig. 1. Before treatment.   Fig. 2. After treatment.

CASE VII.

Fig. 3. Before treatment.   Fig. 4. After treatment.

CASE VIII.

Fig. 5. Before treatment.   Fig. 6. After treatment.

PLATE IV.

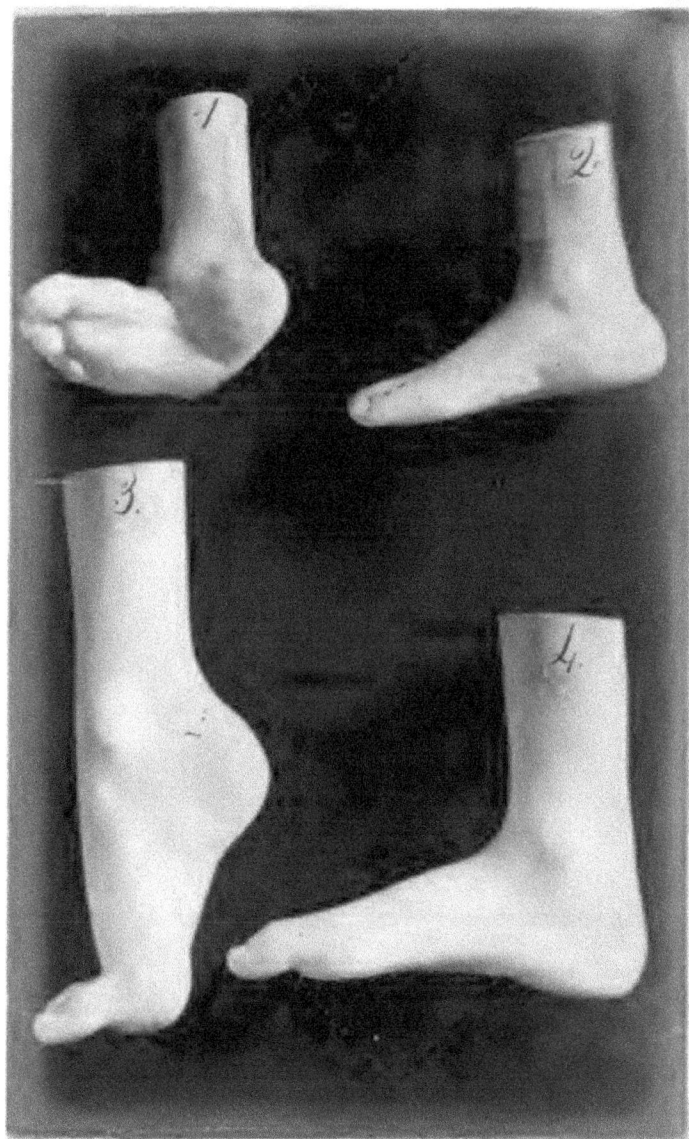

CASE IX.

Fig. 1. Before treatment.   Fig. 2. After treatment.

CASE X.

Fig. 3. Before treatment.   Fig. 4. After treatment.

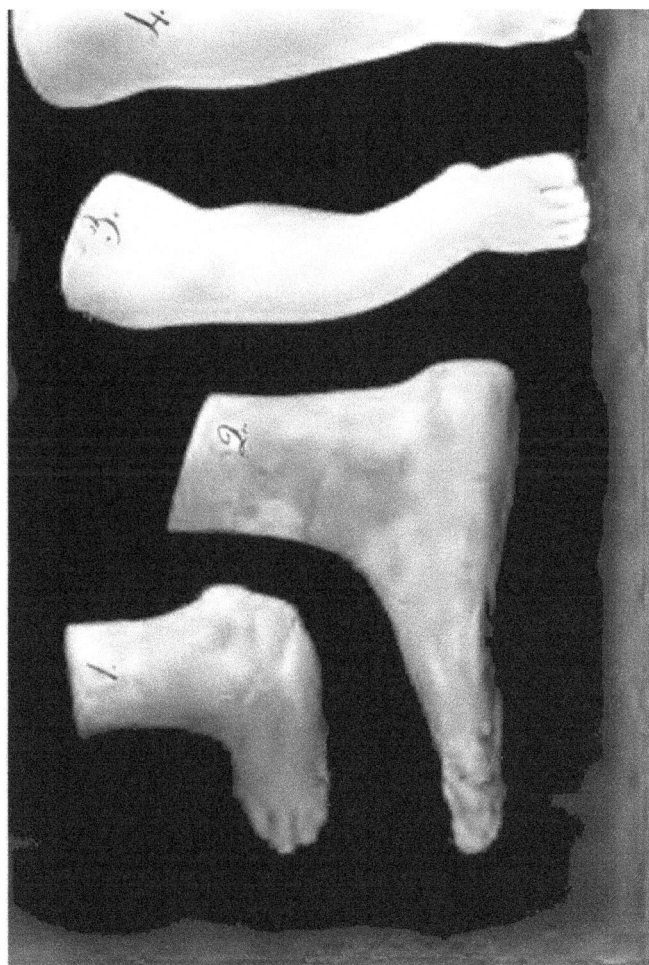

CASE XL.

Fig. 1. Before treatment.   Fig. 2. Two years after treatment.

CASE XLI.

Fig. 3. Before treatment.   Fig. 4. After treatment.

PLATE VI.

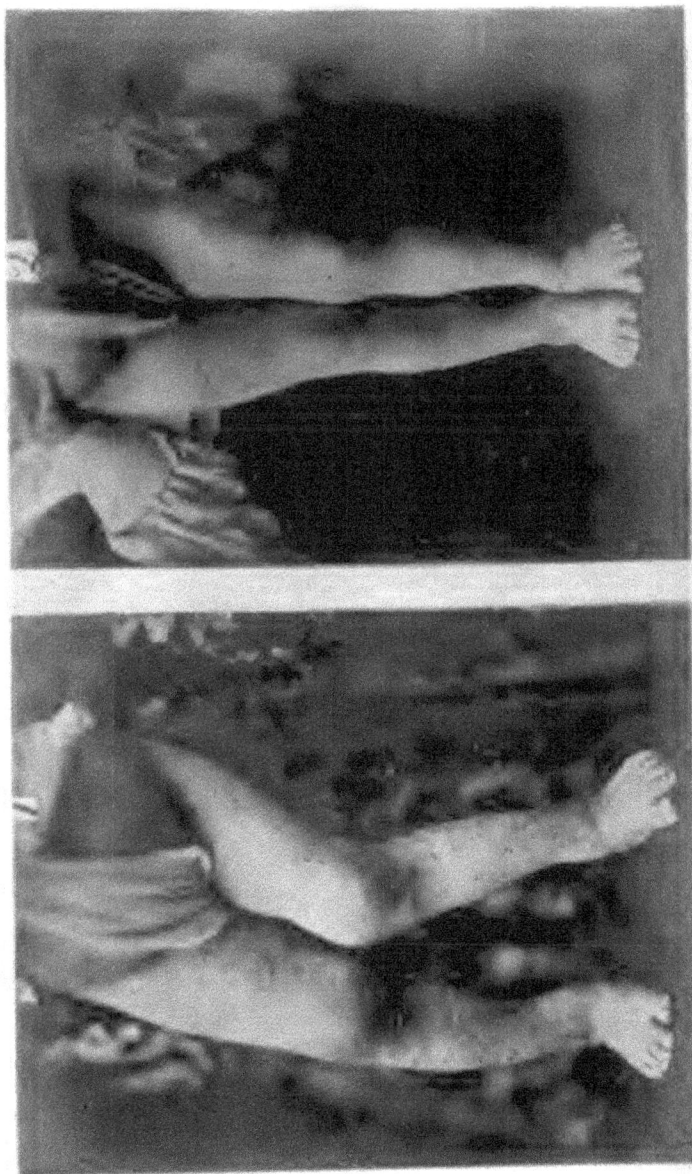

CASE XIII.

Fig. 1. Genu-valgum of right leg, before treatment. Fig. 2. After treatment.

Genu-varum of left leg   "   "

CASE XVI.

LATERAL CURVATURE OF THE SPINE.

Fig. 1. Before treatment.   Fig. 2. After treatment.

PLATE VIII.

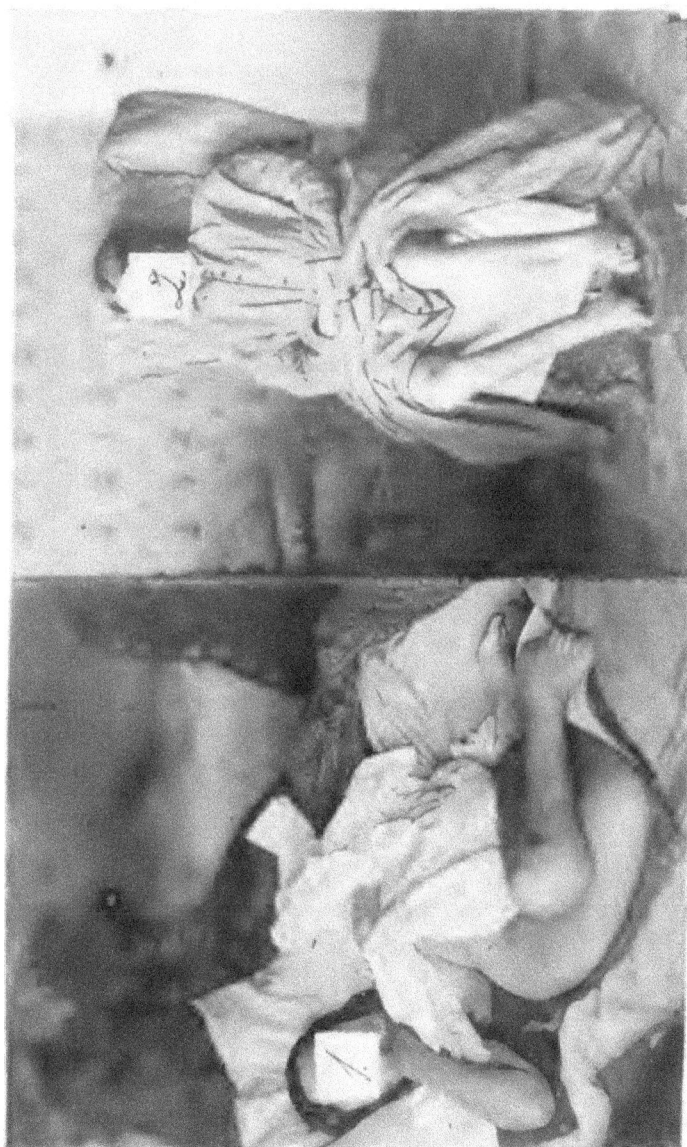

CASE XX.

DISTORTION OF THE SPINE AND LIMBS.

Fig. 1. Before treatment.   Fig. 2. After treatment.

J. J. HAWKE   PHOTO.